Wool

Edited by Rebecca Stefoff

Text © 1990 by Garrett Educational Corporation
First Published in the United States in 1990 by Garrett Educational
Corporation, 130 E. 13th Street, Ada, Ok 74820

First Published in 1989 by A&C Black (Publishers) Limited, London with
the title WOOL © 1989 A&C Black (Publishers) Ltd.

Manufactured in the United States of America.

Library of Congress Cataloging-in-Publication Data

Dixon, Annabelle.
    Wool / Annabelle Dixon ; photographs by Ed Barber.
    p.   cm. - (Threads)
    Includes index.
    Summary: Describes how wool from a sheep is made into a sweater
and explains how to tell the difference between wool and man-made fibers.
Includes activities featuring weaving and dyeing wool.
    ISBN 0-944483-73-9
    1. Wool - Juvenile literature. [1. Wool.] I. Barber, Ed, ill. II. Title.
III. Series.
TS1547.D59   1990
677'.31-dc20
                                          90-40366
                                              CIP
                                              AC

Wool

Annabelle Dixon

Photographs by Ed Barber

Contents

Is it made from wool?   2
Where does wool come from?   8
Wool from sheep   9
Making things from wool   10
How to dye wool   12
Carding wool   14
How to spin wool   16
Knitting   8
Weaving   20
Woolen clothes   24
More thing to do   25
Index   25

GEC GARRETT EDUCATIONAL CORPORATION

Is it made from wool?

These sweaters all look like they are made from wool, but only one is really made from sheep's wool. Can you tell which one it is? *(The answer is on page 25).*

The other sweaters are made from manmade fibers, such as nylon and acrylic. These fibers are made from oil, coal, and wood.

It is difficult to tell if something is made from sheep's wool just by looking at it. But you can often see if your guess was right by reading the labels on clothes or yarn.

Your label says acrylic.

This one's made from wool.

Look for this special symbol on the labels. It is called the woolmark.

The woolmark symbol can only be used if something is made from pure new wool. The same symbol is used all over the world.

PURE NEW WOOL

3

If you can't find a label, you can do some tests to help you tell the difference between manmade fibers and wool.

First you will need a collection. Choose some things you think are made from sheep's wool and some you think are made from manmade fibers.

Here are some ideas

Old clothes
(such as a sweater, a scarf and some gloves)

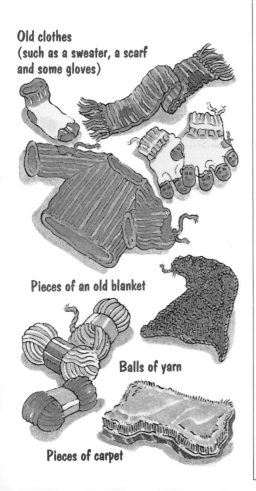

Pieces of an old blanket

Balls of yarn

Pieces of carpet

The feeling test

Feel all the things in your collection and brush them against your cheek.

Wool will feel soft, warm, and bouncy. Manmade fibers will feel flat and cool.

The pulling test

Cut a piece of yarn about as long as this page. Hold it at both ends and stretch it by pulling strongly. Then let go. What happens? Does the yarn spring back?

Wool will spring back even if you do the test many times. But it will probably break more easily than manmade yarns. Manmade yarns will feel like string and so hard. They will not spring back as easily as wool.

The smelling test

Smell all the things in your collection. Some wools smell like fur.

Do they smell different when they are wet? Some people say that the smell of wet wool reminds them of wet dogs. Try doing this test blindfolded.

Turn the page for some more tests.

The dampness test

You will need

One glove you think
is made from wool

One glove you think is
made from manmade fibers

Some water

A paper towel or a paper
napkin—colored one
is best

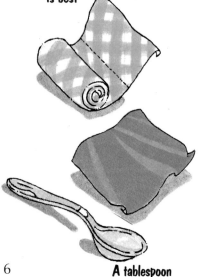

A tablespoon

How to do it

1. Put on the gloves and sprinkle two tablespoons of water over each glove (be sure to get permission first). Rub in the water and wait about five minutes. A wet woolen glove will feel warmer than a wet glove made from manmade fibers.

2. Take off both gloves and place them on the paper towel or napkin (wet side down). A woolen glove will not lose as much water as a glove made from manmade fibers.

The ice cube test

You will need

One glove you think is made from wool

One glove you think is made from manmade fibers

Two small plastic bags

Twist ties

Ice cubes

How to do it

1. Put some ice cubes in each bag and close the bags tightly with the twist ties.

2. Put on the gloves and hold one bag in each hand.

Which ice cubes melt first? Which hand feels cold first? Which glove would you rather wear when it is cold?

Now you know why wool is special. It always feels warm and soft, and it is good at keeping out the cold. Wool can also get quite damp without feeling uncomfortable.

7

Where does wool come from?

Wool does not come only from sheep. All of these animals have the kind of fur we call wool.

▼ We get mohair wool from this kind of goat.

▲ We get angora wool from this kind of rabbit. This one has just had its fur trimmed.

▼ We get vicuna wool from this animal.

All these wools are much more expensive than the wool we get from sheep. So most of our wool comes from sheep.

Wool from sheep

Blackface

Black Welsh Mountain

Dorset Horn

There are many different kinds of sheep. Each kind of sheep has its own kind of wool. Look at these pictures of the wool from these three kinds of sheep. How many differences can you spot?

Making things from wool

How is the wool from a sheep turned into your woolen sweater?

The first step is to cut the wool off the sheep. This feels something like having a haircut; it doesn't hurt the sheep. The wool is cut off in spring or summer when the sheep do not need their thick woolen coats to keep warm. They soon grow new woolen coats.

A sheep's coat is called a fleece. It is cut off with special scissors called shears. Today the shears are usually electric, which makes the job faster and easier, but it's still hard work. It takes about five minutes to shear a sheep. A good sheep shearer can shear about 150 sheep in a day.

The fleeces are weighed and put into sacks, which are called bales. The bales are taken to a factory where they are sorted and washed.

If the fleeces were not washed, your sweater would have 8 tablespoons of oil, a bag of weed seeds, and half a pail of mud in it.

How to dye wool

Before fleeces are made into yarn, they are often dyed. Would you like to try this? (Be sure to get permission first.) Before you add any color, you must do something to the wool that will stop the color from coming out in the wash.

You will need

Some raw sheep's wool or a ball of Aran wool

A deep pan (not aluminum)

A stove or hot plate

Ask an adult to help you with this.

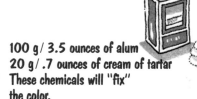

100 g / 3.5 ounces of alum
20 g / .7 ounces of cream of tartar
These chemicals will "fix" the color.
    They are called mordants.

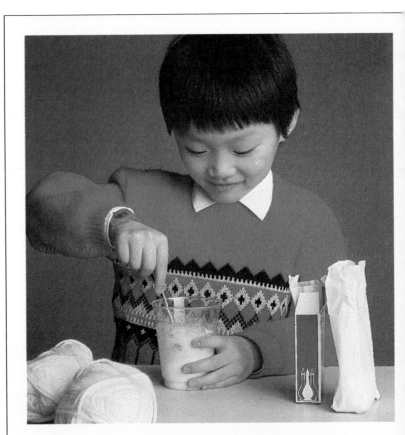

How to do it

Fill the pan with cold water. Mix the alum and cream of tartar with a little water and stir this mixture into the pan. Heat the pan. When the water begins to get warm, add the wool.

Bring the water to a boil and simmer for 45 minutes. Then turn off the heat. When the water is cool, take out the wool. Do not rinse it. Keep it in a plastic bag so it will stay damp until you are ready to dye it.

You can buy packages of chemical dye or
you can make your own dyes from plants.

To make yellow dye you will need

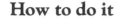

A large paper bag full of brown onion skins (This will make enough dye for about 8 ounces of wool.)

A saucepan

An old spoon

or strainer

A stove or hot plate

How to do it

Bring the onions to a boil in a deep pan full of water and simmer for two hours.

1.   Turn off the heat and fish out the skins with the old spoon or strainer.

2.   Put in the damp wool from the plastic bag and bring the water to a boil again. Simmer for an hour.

Take the wool out of the pan and rinse it in clean, warm water. Leave the wool in a warm place to dry.

13

Carding wool

When the wool has been dyed, it has to be untangled before it can be made into yarn. Remember what you do to your hair after it has been washed? You brush or comb it so that the hairs all lie the same way. The same thing is done to wool. It is called carding the wool.

Carders like these are used to card the wool by hand. In a factory, a machine uses hundreds of small combs to do the same job.

You can try carding wool with an ordinary comb or with two hair brushes. Lay the wool over one brush and pull the other brush over it. You should end up with a little ball of fluffy wool.

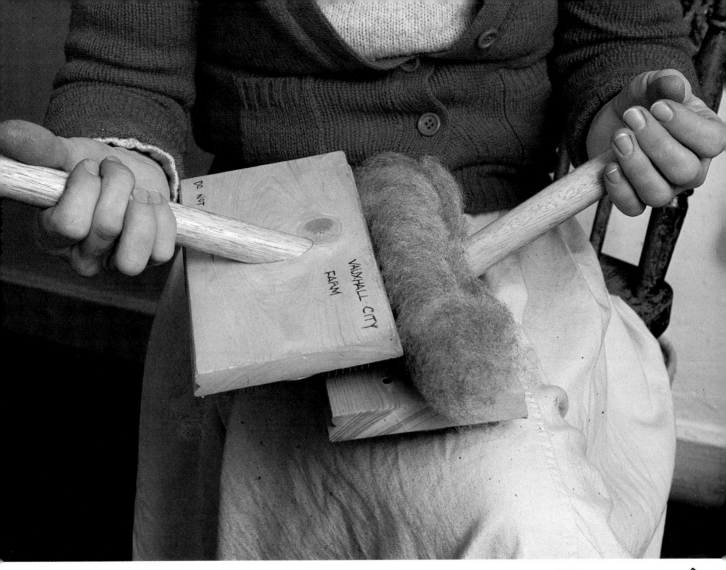

The carded wool is carefully rolled into a loose sausage shape. This is called a rolag or a sliver. The next job is to twist the rolag into a thread.

The picture of wool fibers was taken through a microscope. Can you see that each fiber has a rough, scaly edge? When wool fibers are twisted around each other, these rough scales lock onto each other and stop the fibers from pulling apart. This twisting is called spinning (*Turn the page to see how to spin wool.*)

15

How to spin wool

You can spin wool between your fingers but you will only be able to make lots of short lengths. These are not very useful for making things. To make long threads, you need something called a spindle.

To make a spindle you will need

A lump of modeling clay about as big as a walnut

Animal wool (from a yarn store or drugstore) or a home-made rolag

A plastic foam meat tray

A short knitting needle abou 15-20 cm / 7-9 inches lo

How to do it

1. Place a cup or beaker over the tray and draw around it to make a circle. Cut out the circle.

2. Put the modeling clay about 2 inches from the knob end of the needle and push the circle of plastic foam on top.

3. Twist some of your rolag into a thin thread and tie it between the modeling clay and the knob of the needle. Tie it again about halfway up the needle.

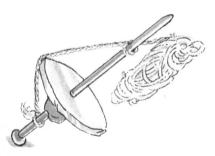

Twist your rolag between your fingers and let your spindle turn around. It should take the yarn with it and start to build up a ball of wool. It will take some practice to get this right.

◀ This is someone using a spindle to spin wool by hand.

▼ A spinning machine in a factory has hundreds of spindles. It works like a hand spindle, but the machines do the twisting.

◀ A spinning wheel is a kind of spindle. Can you figure out why? What job is the foot doing?

17

Knitting

Find an old knitted sweater or scarf and get permission to make a cut in one end so you can pull a loose thread. Watch how the knitting comes undone. Can you see how the thread goes in and out of loops?

Go on pulling until all the knitting has disappeared. You should be left with one long piece of yarn. A whole sweater can be knitted from just one ball of yarn.

When wools of different colors are twisted around knitting needles, it's possible to make lots of different patterns. How many knitted patterns can you see around the edge of this page?

Large knitting needles are used for very thick, chunky wools. The smallest needles make very fine knitting. If you use one needle with a hook, it is called crocheting.

Knitting machines also use needles, but they are different from the needles used for hand knitting. If you look carefully at this photograph, you should be able to see lots of very small needles on the machine.

Weaving

Have you still got the collection you made earlier (see page 4)? If you have a piece of old blanket or scarf, it may not be knitted. Cut off a piece of the blanket or scarf and pull a loose thread from one side. Does it look like this?

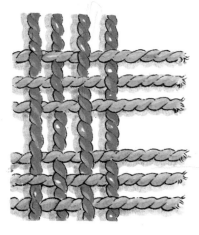

If the material has been woven, you should be able to see one set of threads that go up and down and one set of threads that go from side to side. The up-and-down threads are called warp threads. The side-to-side threads are called weft or woof threads.

Warp threads

Weft threads

When people weave a piece of cloth, they set up the warp threads first. Then they weave the weft threads in and out of the warp threads. In the photograph, can you see that the weft thread is wrapped around a rod in a wooden frame? This tool is called a shuttle.

Try making your own loom

You will need

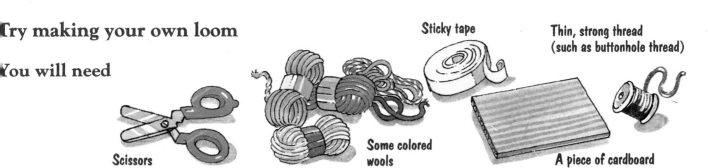

Sticky tape

Thin, strong thread
(such as buttonhole thread)

Scissors

Some colored
wools

A piece of cardboard

How to do it

1. Cut a line of notches in the top and bottom edges of the piece of cardboard. Tape down one end of your warp thread and wind the rest of the thread around the card so it fits through the notches. Pull the thread as tight as you can. When you have finished, tape down the other end of the thread.

2. Now you can choose some different colored yarns to weave in and out of the warp threads. You can use one long thread or lots of short threads.

21

Before you start weaving, it's a good idea to think of a pattern or a picture. Can you see the pattern behind the warp threads on this tapestry?

The pattern helps the weaver remember where to start a different color. Before you wind the warp threads around your loom, you can draw a pattern on the card. Make up some patterns like these.

If you want to make a small mat, weave on one side of your loom.

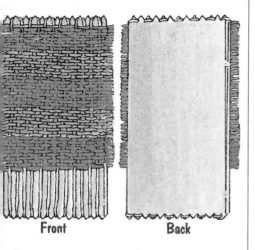

Front   Back

If you want to make a purse, you will have to weave all the way around your loom.

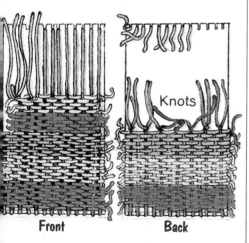

Knots

Front   Back

When you have finished, cut the threads so you can take out the cardboard. Tie a knot in the end of each thread so your weaving won't come undone.

People still weave by hand on simple looms. Can you see how this weaver lifts the warp threads on her loom?

Weaving machines in factories are more complicated than hand looms, but they work in a similar way.   23

Woolen clothes

The next time you wear something made from wool, think about how it was made from a fleece like this one. Can you remember all the things that have to happen to the wool before it can be knitted or woven into clothes?

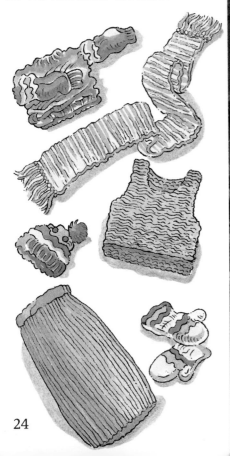

Wool is such a special fiber that even the best manmade fibers cannot copy it yet.

More things to do

1. Find out more about dyeing wool by making different colored dyes from plants, fruits, and vegetables. Try nettles, gooseberries, blackberries, or red cabbage. Make up your own recipes or look in books about dyeing wool.

2. Try out this simple weaving pattern on your cardboard loom. (To see how to make a loom, look at page 21.)
Line 1: over two threads; under one thread.
Line 2: under two threads; over one thread.
Use one color for line 1 and a different color for line 2. Keep this two-line design going until you see a pattern coming. You can make a book from your own weaving patterns and pictures of weaving from magazines or pamphlets.

3. Collect some different colored balls of wool and ask someone to teach you how to knit in plain stitch.
How to make a cushion cover
Knit some simple squares about 10cm / 4 inches long and 10cm / 4 inches wide. Cut off a piece of wool and thread it through a large sewing needle. Sew the edges of the squares together but don't forget to leave one side open so you can put the cushion inside.
How to make a sleeveless top
Use large needles and thick wool to knit eight squares. Make each square about 20cm / 8 inches wide and 20cm / 8 inches long. Sew four of the squares together; then sew the other four squares together. Lay one block of squares on top of the other and sew them together across the shoulders and down the sides. Leave gaps for your head and arms to go through.

4. See if you can find out the answers to these questions. Who were the first people to invent knitting? What does a Guernsey sweater look like? What kind of animal does cashmere wool come from? What patterns are knitted on Fair Isle sweaters? How is felt made?

5. This book has told you about three kinds of sheep. See if you can find out the names of some other kinds of sheep.
How long can you make your list?

Page 2—answer: the sweater on the right is made from wool.

Index

(Numbers in **bold** type are pages that show activities.)

angora wool   8
animal wool   8, 25

carding   **14**, 15
clothes   2, **3**, **4**, 24, **25**
crocheting   19

dyeing wool   **12**, **13**, **25**

fibers   2, **3-7**, **7**, 15, 24
fleece   10, 11, 12, 24

gloves   **6**, **7**

knitting   **18**, 19, **25**

loom   **21**, 22, **23**, **25**

machines   17, 19, 23
manmade fibers   2, **3-7**, **7**, 24
mohair wool   8
mordants   **12**

patterns   18, **22**

rolag   15, **16**, **17**

shearing   10
sheep   8, 9, 10, **25**
shuttle   20
sliver   15
spindle   **16**, **17**
spinning   15, **16**, **17**
spinning wheel   17
sweater   2, 3, 4, 18, 25

tapestry   22
threads   **16**, **17**, **18**, 20, 21, 22, **23**, **25**

vicuna wool   8

warp threads   **20**, **21**, 22
weaving   **20-23**, **25**
weft threads   **20** 21
woolmark   3

yarn   3, **4**, 5, **12**, 18, 21

A female ladybug lays eggs on the underside of a leaf. The eggs look like tiny yellow jelly beans.

A larva grows inside an egg.
The egg becomes darker as the
larva grows bigger.